HOW TO USE AUTOMATIC DIAGNOSTIC SCANNER

From Novice to Expert: The Complete Guide to Understanding and Applying Scan Tool Technology

RORY J. ORTIZ

Copyright © 2024 by RORY J. ORTIZ

All rights reserved

No part of this publication may be reproduced, stored in a retrieval system. or transmitted. in and form or by any means, electronic, mechanical, photocopying, recoraing, or otherwise, without the prior written permission of the author. The information in this ebook is true and complete to the best of our knowledge. All recommendations are made without guarantee on the part of the author or publisher. The author and publisher disclaim any liability in connection with the use of this information.

Table of Contents

Introduction 4
    Overview of Automatic Diagnostic Scanners 7
    Importance of Diagnostic Scanners in Modern Automotive Repair 10
    Objectives of the Guide 14

Understanding Diagnostic Scanners 17
    Types of Diagnostic Scanners 17
    How Diagnostic Scanners Work 20
    Benefits of Using Diagnostic Scanners 23

Preparation for Use 26
    Safety Precautions 26
    Tools and Equipment Needed 28
    Initial Setup and Configuration 30
    Software Installation and Updates 33

Basic Operations 35
    Turning On the Scanner and Vehicle Preparation 35
    Navigating the Interface 37
    Reading and Interpreting Codes 40
    Clearing Codes 43

Advanced Functions 45
    Live Data Monitoring 45
    System Tests (e.g., EVAP system, O2 sensor) 48
    Programming and Coding 50

Special Functions (e.g., DPF regeneration, TPMS reset) 53
Troubleshooting Common Issues 56
  Connectivity Problems 56
  Software and Update Issues 59
  Interpreting Complex Codes 61
  Addressing Inaccurate or Inconsistent Data 64
Maintenance and Care of Your Diagnostic Scanner 67
  Regular Software Updates 67
  Hardware Maintenance 69
  Storage and Handling 72
Conclusion 74

Introduction

In the heart of the bustling city of AutoVille, where the hum of engines and the clank of tools never ceased, lived Sam, a seasoned mechanic with a reputation for diagnosing the most cryptic of car troubles. Sam's secret weapon wasn't just his decades of experience but a gadget no larger than a notebook, yet more powerful than the most sophisticated tool in his garage: an automatic diagnostic scanner.

The story unfolds on a crisp Monday morning when a young woman named Mia drives her stuttering car into Sam's garage. Her car, a modern marvel with more computing power than the first spaceship to the moon, had been giving her sleepless nights. Dashboard lights flickered like a haunted house, and mechanics had quoted her fortunes to even just look at it.

Sam, with a reassuring smile, introduced Mia to his trusted companion—the automatic diagnostic scanner. As he connected the device to her car, he began to narrate a tale not unlike the ones found in a compelling guide he once read, "How to Use an Automatic Diagnostic Scanner."

He spoke of the first chapter, where the guide emphasized the importance of understanding what a diagnostic scanner is and how it can be the bridge between man and modern automobiles. He explained how, just like in the book, he was about to interpret the car's language spoken through error codes.

As the scanner did its magic, Sam shared insights from the guide's chapters on basic operations and advanced functions. He illustrated how live data monitoring allowed him to see in real time what troubled Mia's car. With each passing moment, Mia saw not just a mechanic at work but a craftsman wielding knowledge as his tool.

Sam's narrative took a turn towards the practical, echoing a chapter from the book on troubleshooting common issues. He demonstrated how the scanner pinpointed the problem to a faulty sensor, saving hours of manual diagnosis and unnecessary repairs. Mia's anxiety turned to awe as she realized the value of understanding and utilizing such a tool.

By the time the car was fixed, Mia was not just relieved but enlightened. Sam had shared tales of the guide's sections on emerging technologies and how owning a diagnostic scanner is not just about fixing problems today but staying ahead in the future. He spoke of electric vehicles, telematics, and how the book covered these topics in depth, preparing its readers for the automotive evolution.

As Mia drove away, satisfied and curious, Sam handed her a flyer for the guidebook, "How to Use an Automatic Diagnostic Scanner." He explained that the book was more than just a manual; it was a gateway to empowering oneself in an era where cars speak a digital language. For anyone who owned a car or aspired to understand the heartbeats of these mechanical beasts, this guide was the key.

The story of Sam and Mia is but one of the many that unfolded in AutoVille, a testament to the transformative power of knowledge. The guidebook Sam advocated didn't just teach people to use a tool; it inspired them to be more, to demystify the complexities of modern vehicles, and to save not just money but time and peace of mind.

And so, in every home, garage, and glove compartment where the guide found its place, it whispered the promise of confidence, the assurance that in the world of automotive mysteries, you too could be a detective, if only you knew how to use an automatic diagnostic scanner.

Overview of Automatic Diagnostic Scanners

Automatic diagnostic scanners have revolutionized the way we understand, diagnose, and fix vehicles. These compact yet powerful devices serve as the crucial link between the modern mechanic and the increasingly complex systems of today's automobiles. In the context of "How to Use an Automatic Diagnostic Scanner," understanding these tools' capabilities, applications, and importance is essential for anyone looking to navigate the automotive repair world effectively.

What are Automatic Diagnostic Scanners?

Automatic diagnostic scanners are electronic tools that interface with a vehicle's on-board diagnostics system (OBD). Every vehicle manufactured since the early 1990s is equipped with an OBD system, which monitors various sensors and systems within the vehicle to ensure everything is operating correctly. When something goes awry, the OBD system generates diagnostic trouble codes (DTCs), which can be accessed using a diagnostic scanner.

Capabilities and Features

These scanners range from basic code readers that can only read and clear codes, to advanced devices capable of a wide array of functions, including:

- ✓ Reading real-time data from the vehicle's sensors
- ✓ Performing system tests (e.g., fuel system test, EGR system)
- ✓ Programming new parts
- ✓ Adjusting vehicle settings

Advanced scanners can provide detailed descriptions of DTCs, suggest potential fixes, and guide users through complex troubleshooting processes. They can also display live data in real-time, providing invaluable insights into the vehicle's operation, far beyond what is possible with traditional diagnostic methods.

Importance in Modern Automotive Repair

The advent of electronic control units (ECUs) in vehicles has made them more efficient and reliable but also more complex. The diagnostic scanner becomes an indispensable tool in this context, allowing mechanics and enthusiasts to interact with these sophisticated systems efficiently. It empowers users to quickly identify issues, understand their severity, and decide on the appropriate course of action, saving both time and money.

Using an Automatic Diagnostic Scanner

Using an automatic diagnostic scanner typically involves several steps:
1. Connection: The scanner is plugged into the OBD-II port, usually located under the dashboard.

2. Initial Scan: The user performs an initial scan to read any stored DTCs.

3. Interpretation: The scanner may provide information on the code and possible causes. Further research or a detailed guide can help interpret more complex codes.

4. Troubleshooting and Repair: Based on the codes and additional diagnostic information, the user can troubleshoot and perform necessary repairs.

5. Clearing Codes: After repairs, the scanner can clear the codes and reset the check engine light.

Educational Value

For enthusiasts and professionals alike, understanding how to use an automatic diagnostic scanner is invaluable. It not only facilitates efficient repairs but also provides a deeper understanding of automotive systems. Guides and manuals, such as "How to Use an Automatic Diagnostic Scanner," offer comprehensive insights into leveraging these tools effectively, covering everything from basic operations to advanced diagnostics and troubleshooting.

Automatic diagnostic scanners represent a critical advancement in automotive technology, bridging the gap between complex vehicle systems and the individuals tasked with their maintenance and repair. By demystifying the process of diagnosing and fixing modern vehicles, these scanners empower users to take control of automotive care, making them an essential topic in any automotive education resource.

Importance of Diagnostic Scanners in Modern Automotive Repair

The landscape of automotive repair has undergone a monumental shift with the advent of digital technology. At the forefront of this transformation is the use of automatic diagnostic scanners, an indispensable tool in the modern mechanic's arsenal. Understanding and effectively using these scanners is not just an enhancement but a necessity for anyone involved in automotive repair, maintenance, or even ownership. The significance of diagnostic scanners in today's automotive repair environment is multifaceted, encompassing efficiency, accuracy, cost-effectiveness, and the advancement towards a more informed and tech-savvy automotive culture.

Efficiency and Time-Saving

Before the era of diagnostic scanners, pinpointing problems in vehicles was often akin to looking for a needle in a haystack. Mechanics had to rely on manual inspection and trial-and-error methods, which were not only time-consuming but also prone to inaccuracies. With the introduction of automatic diagnostic scanners, the time to diagnose issues has drastically reduced. These scanners can swiftly read error codes generated by a vehicle's onboard computer system, directly pointing to the malfunctioning component or system. This rapid diagnosis is crucial in a fast-paced world, ensuring that vehicles spend less time in the shop and more on the road.

Enhanced Accuracy

Modern vehicles are complex machines, with an intricate network of electronic systems and sensors. The precision that automatic diagnostic scanners bring to the table is unparalleled. They provide detailed information on the health of various components, from the engine and transmission to the exhaust system and beyond. This accuracy eliminates guesswork, allowing for direct and effective repairs. By detailing specific error codes and sometimes even suggesting potential fixes, these scanners ensure that the exact issue is addressed, enhancing the quality of repairs and minimizing the risk of misdiagnosis.

Cost-Effectiveness

The precision and efficiency of automatic diagnostic scanners translate directly into cost savings. For repair shops, the ability to quickly diagnose and address issues means a higher turnover of repair jobs and reduced labor costs. For vehicle owners, it means paying for the exact repair needed without the added expenses of unnecessary parts or extended labor hours. Furthermore, by identifying potential problems early, these scanners can help avoid more severe and costly repairs down the line, promoting regular maintenance and prolonging vehicle lifespan.

Empowering Owners and Technicians

Knowledge is power, and in the context of automotive repair, this couldn't be truer. Automatic diagnostic scanners demystify the inner workings of vehicles, making the information accessible not just to technicians but to vehicle owners as well. This empowerment fosters a better understanding of how vehicles operate and the importance of regular maintenance. For aspiring mechanics and DIY enthusiasts, mastering the use of these scanners through guides like "How to Use an Automatic Diagnostic Scanner" is a stepping stone into the broader world of automotive repair and maintenance, equipping them with the skills and knowledge to tackle challenges confidently.

Adapting to the Future

The automotive industry is on the brink of a technological revolution, with electric vehicles (EVs) and advanced driver-assistance systems (ADAS) becoming increasingly prevalent. The role of diagnostic scanners is set to become even more critical as these new technologies demand specialized diagnostics and understanding. Staying abreast of how to use and interpret the readings from automatic diagnostic scanners is crucial for anyone looking to remain relevant in the automotive repair field.

In conclusion, the importance of automatic diagnostic scanners in modern automotive repair cannot be overstated. They are the key to unlocking efficiency, accuracy, cost savings, and empowerment in the face of evolving vehicle technologies. For mechanics, technicians, and vehicle enthusiasts alike, acquiring a deep understanding of these tools through comprehensive guides and hands-on experience is indispensable. As vehicles continue to advance, so too must our tools and knowledge, ensuring that we can keep pace with the innovations that drive the automotive world forward.

Objectives of the Guide

The guidebook on "How to Use an Automatic Diagnostic Scanner" is crafted with precision to serve a multifaceted purpose, aimed at demystifying the complex nature of automotive diagnostics for a diverse audience. Whether you're a professional mechanic, an automotive enthusiast, or a car owner with no technical background, this guide seeks to empower you with the knowledge and skills needed to utilize diagnostic scanners efficiently. Here are the detailed objectives of the guide:

1. Educate on the Basics of Diagnostic Scanners

The guide begins with foundational knowledge, introducing readers to what diagnostic scanners are, the different types available (from basic code readers to advanced professional scanners), and how they interface with a vehicle's onboard diagnostics system (OBD). Understanding these basics is crucial for anyone looking to effectively use these tools.

2. Simplify Complex Diagnostic Processes

One of the main objectives is to break down complex diagnostic processes into simple, easy-to-follow steps. This demystification enables users to not only operate the scanners but also to interpret the data they provide, translating error codes and warnings into understandable language.

3. Enhance Troubleshooting Skills

The guide aims to refine the troubleshooting skills of its readers. By providing case studies, common scenarios, and how to approach various faults, it ensures that users can approach vehicle diagnostics with confidence, reducing reliance on costly professional diagnostics for every issue.

4. Promote Informed Decision Making

With the knowledge gained from this guide, readers will be able to make more informed decisions regarding repairs and maintenance. Understanding the specific issues affecting a vehicle allows for better communication with repair professionals and ensures that readers are not easily misled.

5. Foster a Do-It-Yourself (DIY) Culture

Encouraging a DIY culture among car owners and enthusiasts is a significant objective. The guide equips readers with the confidence to undertake minor repairs and maintenance tasks themselves, promoting a hands-on approach to vehicle care and potentially saving considerable amounts of money over time.

6. Update Readers on Latest Technologies

As automotive technology evolves, so do diagnostic tools. The guide includes information on the latest advancements in diagnostic scanners, including wireless and Bluetooth options, and how they can be used with modern vehicles, ensuring readers are up-to-date.

7. Provide Safety and Legal Guidelines

Using diagnostic scanners involves certain legal and safety considerations. The guide emphasizes the importance of understanding these aspects, offering advice on how to use scanners without violating privacy laws or vehicle warranty terms and ensuring personal and vehicle safety during diagnostics.

8. Support Continuous Learning

Finally, the guide is designed as a springboard for continuous learning. It includes resources for further study, such as websites, forums, and additional reading materials. This objective recognizes that the field of automotive diagnostics is ever-changing, and staying informed is key to maintaining proficiency.

In essence, the guide on "How to Use an Automatic Diagnostic Scanner" is a comprehensive resource that aims to equip readers with the knowledge and skills necessary to confidently and efficiently diagnose and troubleshoot vehicle issues. Its objectives reflect a commitment to education, empowerment, and continuous improvement in the field of automotive diagnostics.

Understanding Diagnostic Scanners

Types of Diagnostic Scanners

In the realm of automotive diagnostics, the tools at our disposal have evolved dramatically, giving rise to a variety of diagnostic scanners each designed to cater to different needs, expertise levels, and budgets. The understanding of these tools is fundamental for anyone looking to delve into the mechanics of automobiles, whether for professional repair work or personal maintenance.

Basic Code Readers are the simplest form of diagnostic tools available in the market. They are designed to do exactly what their name suggests: read and clear basic error codes from a vehicle's onboard diagnostics system. These devices are typically compact, easy to use, and affordable, making them an excellent entry point for car owners who wish to tackle basic diagnostic tasks themselves. However, their functionality is limited to reading generic fault codes, with no capability to access manufacturer-specific codes or provide detailed information about the source of the problem.

Professional Scanners are at the other end of the spectrum, offering a comprehensive suite of diagnostic features far beyond what basic code readers can provide. These devices can access a wide range of data, including manufacturer-specific codes, live data streaming, and advanced troubleshooting information. Professional scanners often come with extensive databases and support materials to assist in diagnosing and solving complex automotive problems. Designed for mechanics and serious automotive enthusiasts, these scanners are equipped with features for programming and calibration tasks, making them indispensable in a professional setting. The complexity and cost of these tools, however, can be a barrier for casual users.

Bluetooth and Wireless Scanners represent the modern evolution of diagnostic tools, offering a blend of convenience and functionality that caters to the needs of today's tech-savvy users. These scanners connect to a vehicle's OBD port via a dongle and transmit data wirelessly to a smartphone or tablet, using an app to interpret and display the information. This setup allows for unparalleled ease of use and mobility, enabling users to perform diagnostics and monitor their vehicle's performance in real-time, without the need for bulky hardware. While these scanners vary in capabilities, many offer a range of features that rival those of professional scanners, including access to live data and manufacturer-specific codes. The software-based nature of these tools means they can be regularly updated with new features and support for newer vehicle models, making them a versatile and future-proof choice for both casual and professional users alike.

Each type of diagnostic scanner has its place in the toolkit of anyone interested in automotive diagnostics, from the casual car owner to the professional mechanic. Understanding the strengths and limitations of each is crucial in selecting the right tool for your needs, ensuring that you can not only interpret what your vehicle is telling you but also make informed decisions about maintenance and repairs.

How Diagnostic Scanners Work

Diagnostic scanners are pivotal tools in the realm of automotive repair and maintenance, enabling both professionals and enthusiasts to interact with the increasingly complex electronic systems found in modern vehicles. These devices work by connecting to a vehicle's onboard diagnostics port (OBD-II), a standard feature in cars manufactured from 1996 onwards in the United States. This port serves as a gateway to the car's electronic control units (ECUs), which monitor and manage various aspects of the vehicle's performance, including engine and transmission operations, emissions systems, and more.

Once connected, a diagnostic scanner communicates with the vehicle's ECUs to retrieve diagnostic trouble codes (DTCs). These codes are generated and stored by the ECUs when they detect irregularities or failures in the vehicle's systems. Each code corresponds to a specific issue, ranging from minor sensor malfunctions to potential engine problems. The ability of scanners to read these codes is what makes them invaluable; they translate complex electronic signals into actionable information.

Basic code readers can only read and clear codes, offering a straightforward indication of where a problem might lie. However, more advanced scanners provide a wealth of information beyond simple codes. They can display live data from the sensors and actuators as the vehicle operates, allowing for real-time monitoring of engine performance, fuel efficiency, and other critical parameters. This capability is essential for understanding how intermittent problems occur and for verifying that repairs have effectively addressed the issue.

Furthermore, professional-grade diagnostic scanners offer functionalities such as actuation tests, system calibration, and even programming new components. These advanced operations enable mechanics and technicians to interact with the vehicle's systems directly, performing detailed diagnostics and adjustments that were once only possible at dealership service centers.

Diagnostic scanners also vary in their method of connectivity and user interface. While traditional models require a physical connection via a cable, newer versions use Bluetooth or WiFi to connect wirelessly to a smartphone or tablet app. This flexibility allows for a more user-friendly experience, as these apps can provide guided diagnostics, definitions for trouble codes, and even repair suggestions.

An essential aspect of how diagnostic scanners work is their reliance on software, both within the scanner itself and the vehicle's ECUs. Regular updates to the scanner's software are necessary to ensure compatibility with the latest vehicle models and their diagnostic protocols. Similarly, understanding the vehicle's software version can be crucial when diagnosing issues or performing updates to the vehicle's firmware.

In essence, diagnostic scanners serve as the bridge between the user and the sophisticated electronics of modern vehicles. By translating diagnostic codes and providing real-time data and advanced functionality, these tools demystify the process of automotive troubleshooting and repair, making them indispensable to anyone seeking to understand or work on modern vehicles.

Benefits of Using Diagnostic Scanners

The advent of diagnostic scanners revolutionized automotive maintenance and repair, offering a plethora of benefits that cater to both professional mechanics and car enthusiasts alike. These devices facilitate a direct dialogue with a vehicle's onboard computer system, allowing for the swift identification of both present and potential issues. This immediacy in diagnostics not only saves time but significantly reduces the guesswork and manual inspection previously required, leading to more accurate and cost-effective repairs.

Diagnostic scanners are pivotal in maintaining vehicle health, offering the ability to monitor live data from various sensors in real time. This capability means that users can detect anomalies before they evolve into serious problems, promoting preventative maintenance and potentially extending the lifespan of the vehicle. Moreover, the ability to clear error codes after issues have been resolved streamlines the repair process, ensuring that only current problems are being addressed.

In an era where vehicles are increasingly complex, with intricate electronic systems and computerized controls, these scanners serve as a critical tool in understanding and maintaining modern automobiles. They empower owners and technicians to perform a wide range of diagnostics and adjustments that were once the exclusive domain of specialized dealerships. This democratization of vehicle maintenance not only helps in reducing dependency on expensive professional services but also enhances the user's knowledge and understanding of their vehicle.

Furthermore, the use of diagnostic scanners significantly enhances the efficiency of repairs and maintenance. By pinpointing the exact source of a problem, they eliminate the need for extensive manual testing and trial-and-error methods, saving both time and resources. This efficiency is especially valuable in a professional setting, where time is of the essence, and accuracy is paramount.

The environmental benefits of using diagnostic scanners are also noteworthy. By ensuring that vehicles operate at optimal efficiency, these tools play a crucial role in reducing emissions. Cars in poor condition tend to emit more pollutants, and through accurate diagnostics, vehicles can be maintained in a state that minimizes their environmental impact.

Diagnostic scanners also offer educational benefits, providing users with detailed insights into the functioning and health of their vehicles. This knowledge fosters a deeper connection between drivers and their cars, encouraging a more responsible and informed approach to vehicle use and maintenance.

Lastly, the evolution of diagnostic scanners to include features such as wireless connectivity and compatibility with smartphones and tablets has further expanded their accessibility and convenience. This technological advancement allows users to perform diagnostics and receive updates in real-time, enhancing the user experience and expanding the possibilities of what can be achieved through their use.

The benefits of using diagnostic scanners are extensive and multifaceted, encompassing improvements in efficiency, cost savings, educational value, environmental impact, and beyond. As vehicles continue to evolve, these tools will undoubtedly play an increasingly vital role in the world of automotive maintenance and repair, making them an indispensable asset for anyone seeking to maintain or repair a vehicle in today's technologically advanced landscape.

Preparation for Use

Safety Precautions

When preparing to use an automatic diagnostic scanner, prioritizing safety is crucial to prevent accidents, protect the vehicle's electronic systems, and ensure your well-being. First and foremost, always consult the vehicle and scanner's user manuals for specific safety instructions and compatibility information. Understanding these guidelines is essential before connecting the scanner to a vehicle.

Ensure the vehicle is parked on a stable, level surface and that the ignition is off before connecting the diagnostic scanner. This reduces the risk of accidental movement and electrical issues. When connecting the scanner, be mindful of the vehicle's battery; a weak or unstable battery can lead to inaccurate diagnostic readings and may even damage the scanner.

It's important to wear appropriate personal protective equipment (PPE), such as safety glasses and gloves, especially when working near the engine compartment. This protection minimizes the risk of injuries from hot surfaces, moving parts, or electrical shocks.

Maintain a clean work environment free of spills, grease, and clutter. This minimizes the risk of slips, trips, and falls, as well as fire hazards. Also, ensure that the diagnostic scanner and its cables are in good condition, without frayed wires or broken connectors, to prevent electrical shocks and ensure accurate diagnostics.

When the diagnostic scanner is in use, avoid touching electrical components or connectors with bare hands, as this can lead to accidental shocks or may interfere with the vehicle's electrical system. If the vehicle needs to be running for a diagnostic test, ensure the area is well-ventilated to avoid the accumulation of exhaust fumes, which can be harmful to your health.

Be cautious not to force connectors into ports, as this can damage both the diagnostic tool and the vehicle's diagnostic port. Always connect and disconnect the scanner and its components gently.

Keep in mind that while diagnostic scanners are powerful tools, they are also sensitive electronic devices. Exposure to extreme temperatures, moisture, or direct sunlight can damage the scanner, affecting its functionality and lifespan.

Lastly, respect the complexity of vehicle electronic systems. If you encounter complex codes or issues beyond your understanding, consult a professional. Attempting to fix issues without proper knowledge can lead to further damage to the vehicle or personal injury.

By adhering to these safety precautions, users can safely and effectively use an automatic diagnostic scanner, ensuring accurate diagnostics while protecting both themselves and the vehicle from potential harm.

Tools and Equipment Needed

To embark on the journey of using an automatic diagnostic scanner effectively, having the right set of tools and equipment is paramount. This ensures not only the proper diagnosis of vehicle issues but also the safety and efficiency of the process.

At the core of the necessary equipment is the diagnostic scanner itself. Selection between a basic code reader and a more advanced scanner should be based on the user's needs, considering features like live data streaming, the ability to read manufacturer-specific codes, and compatibility with various vehicle makes and models.

Software plays a critical role in the functionality of diagnostic scanners. Users should ensure they have access to the latest version of the scanner's software, which may involve subscriptions for updates. This software is crucial for reading the latest codes and accessing new features that enhance diagnostic accuracy.

A laptop or a mobile device often becomes necessary when using advanced scanners or when needing to research codes and solutions online. For wireless or Bluetooth scanners, having a compatible device with the requisite apps installed is essential for seamless operation.

Vehicle service manuals, either in physical form or online, are invaluable in understanding the specific diagnostic codes and troubleshooting steps for particular vehicle models. These manuals provide detailed information on manufacturer-specific codes and repair procedures, essential for accurate diagnostics and repairs.

A reliable internet connection becomes indispensable, especially when using online resources for troubleshooting, downloading software updates, or consulting online forums for advice.

For vehicles that require a physical connection, an OBD-II connector cable is necessary. This cable connects the scanner to the vehicle's diagnostic port, allowing for communication between the scanner and the vehicle's onboard computer.

Personal protective equipment (PPE) such as gloves and safety glasses is recommended to protect against potential hazards like electrical shocks or battery acid. Working with vehicles can expose users to various risks, making PPE an essential aspect of the preparatory phase.

Lastly, a notepad and pen for taking notes or a digital equivalent can be incredibly useful. Recording error codes, symptoms, and any observations during the diagnostic process can aid in troubleshooting and keeping track of vehicle health over time.

By assembling these tools and equipment, users can prepare themselves for a comprehensive diagnostic experience, ensuring they are well-equipped to tackle a wide range of vehicle issues with confidence and precision.

Initial Setup and Configuration

Setting up and configuring an automatic diagnostic scanner is a critical first step in utilizing this powerful tool effectively for vehicle diagnostics. Before diving into the complexities of automotive systems, ensuring that the scanner is properly prepared can make the diagnostic process smoother and more efficient.

The initial setup begins with unpacking the scanner from its packaging, inspecting it for any physical damage that may have occurred during shipping. After confirming the condition, the next step involves charging the scanner if it comes with a rechargeable battery, or installing batteries if it's battery-operated. It's essential to ensure that the device has enough power to perform diagnostics without interruptions.

Once the power setup is complete, the next phase is installing any necessary software or applications that come with the scanner. This might involve connecting the scanner to a computer or a smartphone, depending on the model and capabilities of the device. Downloading and installing the latest software or app version is crucial, as manufacturers often release updates that enhance the tool's functionality and compatibility with newer vehicle models.

After installing the software, the scanner typically requires registration or activation. This process may involve entering a serial number or a product key and possibly creating an account with the manufacturer. This step is vital for accessing support, software updates, and additional features provided by the manufacturer.

Configuration settings are the next focus area. This involves setting up preferences such as the language, units of measurement (e.g., Celsius vs. Fahrenheit, kilometers vs. miles), and any wireless or Bluetooth connectivity options if the scanner supports such features. For scanners that connect to a vehicle using a wired connection, ensure that the OBD-II connector is in good condition and securely attach it to the scanner.

Connecting the scanner to the vehicle for the first time is a critical part of the setup process. Locate the OBD-II port in the vehicle, which is typically found under the dashboard near the steering column, though the exact location can vary depending on the vehicle make and model. Once located, gently plug in the scanner's connector to the port, ensuring a secure connection.

After the physical connection is established, the scanner will usually power on automatically. It may then undergo a self-test or prompt the user to select the vehicle make and model from its database to ensure accurate diagnostics. This step is crucial for tailoring the scanner's operation to the specific vehicle, allowing for precise readings and diagnostics.

The final step in the initial setup involves performing a preliminary scan of the vehicle. This initial scan checks for any existing error codes or issues and helps familiarize the user with the scanner's interface and functions. It's an opportunity to ensure that the scanner is working correctly and is properly communicating with the vehicle's onboard diagnostics system.

Throughout this process, referring to the scanner's user manual is essential for understanding specific instructions, features, and troubleshooting tips related to the model in use. Proper initial setup and configuration set the stage for effective and efficient use of the automatic diagnostic scanner, enabling users to diagnose and address vehicle issues with confidence.

Software Installation and Updates

When venturing into the realm of using an automatic diagnostic scanner, one crucial step is ensuring the device's software is correctly installed and regularly updated. This process is pivotal because the effectiveness of your diagnostic scanner heavily relies on the software that powers it. Initially, the installation involves connecting your scanner to a computer or, in some cases, directly to the internet if it supports Wi-Fi capabilities. This connection allows you to download the necessary software or updates directly from the manufacturer's website or a provided software platform.

Manufacturers often release software updates to enhance the functionality of the scanner, add new features, or expand the database of vehicle codes and troubleshooting information. These updates are essential for keeping your scanner effective across a wide range of vehicle makes and models, including the latest releases. It's also important to note that updates can fix bugs or issues identified in the software, ensuring your device operates smoothly and reliably.

Setting up the scanner usually involves a straightforward installation process, guided by on-screen instructions or a step-by-step guide found in the user manual. This process might include registering the product online, which can provide additional benefits such as access to technical support and notifications for upcoming updates.

Regularly checking for software updates is a practice that cannot be overstated. Depending on the model of your scanner, this might be a manual process where you periodically connect your scanner to check for updates, or it could be automated through Wi-Fi connectivity, where the scanner notifies you of available updates. Regardless of the method, staying current is key to ensuring the scanner's accuracy and reliability.

Moreover, the installation of software and updates often requires a stable internet connection. This requirement highlights the importance of having reliable internet access, especially for scanners that update automatically. In scenarios where a scanner is used in a professional setting, like an auto repair shop, ensuring that the scanner has consistent internet access can save time and prevent delays in diagnostics and repairs.

Lastly, understanding the software interface and its features plays a crucial role in maximizing the scanner's potential. After installation and updates, familiarizing yourself with the software's layout, settings, and how to navigate through its options will enable you to perform diagnostics more efficiently. This knowledge base extends to understanding how to interpret updates, recognizing new features, and applying them to diagnostic procedures.

In summary, the installation and regular updating of your diagnostic scanner's software are foundational steps in preparing to use the device. These actions ensure the scanner remains a reliable tool for diagnosing and understanding a wide array of vehicle issues, making them indispensable practices for anyone looking to leverage the full capabilities of their diagnostic scanner.

Basic Operations

Turning On the Scanner and Vehicle Preparation

Turning on the scanner and preparing the vehicle properly are fundamental steps in the diagnostic process, crucial for both safety and accuracy. Before engaging with an automatic diagnostic scanner, ensure the vehicle is parked on a level surface and the ignition is off. This basic preparation helps prevent accidents and ensures that the vehicle's systems are in a stable state before diagnostics begin.

First, locate the vehicle's OBD-II port, typically found under the dashboard near the steering column, though its exact location can vary depending on the vehicle. Ensure the port is clean and free of debris to facilitate a good connection. Next, take the diagnostic scanner and connect it to the OBD-II port. With the connection secure, the scanner can now be turned on. In most cases, the device will power up automatically upon connection to the port, thanks to the vehicle's battery.

After turning on the scanner, it's crucial to turn the vehicle's ignition to the "on" position without starting the engine. This action powers up the vehicle's electronic systems, allowing the scanner to communicate with the vehicle's computer. Some vehicles may require the engine to be running for a full diagnostic check. Therefore, always refer to the scanner's manual and the vehicle's documentation for specific instructions.

Once the ignition is on, the scanner will typically perform an initial self-check, indicated by a series of lights or messages on its display. After this self-check, the device should display the main menu or interface, signaling it is ready for use. From here, you can navigate through the scanner's options to select the desired diagnostic operation, such as reading error codes or viewing live data.

It's also important to note that proper vehicle preparation extends beyond just the physical setup. Ensuring the vehicle's battery is fully charged can prevent data loss or communication errors during the diagnostic process. Additionally, removing any non-essential electronic devices that may interfere with the vehicle's electrical system can further ensure an accurate diagnosis.

In the context of basic operations, turning on the scanner and preparing the vehicle correctly are pivotal first steps that set the stage for successful diagnostic sessions. These procedures ensure that both the tool and the vehicle are ready for a thorough examination, paving the way for accurate troubleshooting and efficient vehicle maintenance. Following these steps carefully not only enhances the diagnostic accuracy but also contributes significantly to the overall safety of the diagnostic process.

Navigating the Interface

Navigating the interface of an automatic diagnostic scanner is a critical skill for efficiently diagnosing and understanding vehicle issues. The interface, which serves as the primary means of interaction between the user and the scanner's vast array of functions, can vary significantly across different models and brands. However, most follow a structured design philosophy that prioritizes ease of use and accessibility.

Upon powering up a diagnostic scanner, users typically encounter a main menu that offers a selection of options, including reading and clearing error codes, viewing live data, and accessing vehicle-specific information. The key to effective navigation is understanding these main menu options and how they relate to the scanner's capabilities.

Selecting the option to read error codes initiates a scan of the vehicle's onboard diagnostics system. This process, which is usually straightforward, involves the scanner communicating with the vehicle's computer to retrieve any stored codes that indicate malfunctions or issues. Once retrieved, these codes are displayed on the screen, often alongside brief descriptions. Some scanners offer additional details or suggest potential fixes, guiding users towards the next steps in the diagnostic process.

Clearing codes, another fundamental operation, is typically done after addressing the issue that triggered the code in the first place. This function resets the vehicle's diagnostic system, ensuring that only current, unresolved issues are flagged. It's crucial to understand that clearing codes without resolving the underlying problems is a temporary solution; the codes will likely reappear if the issues persist.

Viewing live data is an advanced feature that allows users to monitor the real-time performance of various vehicle sensors and components. This function displays a wealth of information, from engine temperature and RPM to oxygen sensor readings, providing insights into the vehicle's operational state. Navigating this section involves scrolling through data streams and selecting specific parameters to monitor. Some scanners allow users to graph multiple data points simultaneously, offering a more comprehensive view of the vehicle's performance.

Vehicle-specific information, another useful feature, offers access to a range of data tailored to the make, model, and year of the vehicle. This can include maintenance schedules, technical specifications, and even troubleshooting guides. Accessing this information requires navigating through menus to input or select the appropriate vehicle details.

Throughout the navigation process, users should make use of any on-screen instructions or help functions provided by the scanner. Many models feature a help or info button that, when pressed, displays detailed explanations of the current screen or function. This feature is invaluable for understanding complex features or deciphering cryptic error codes.

Finally, effective navigation depends on familiarity with the scanner's physical interface. Buttons, knobs, or touchscreens are the main controls, and their responsiveness and layout can significantly impact the user experience. Practicing with these controls, exploring menu structures, and experimenting with different functions are essential steps in becoming proficient with an automatic diagnostic scanner.

In essence, navigating the interface of an automatic diagnostic scanner is about understanding its structure, utilizing its features to access and interpret diagnostic data, and leveraging any available resources to enhance the diagnostic process. Mastery of this aspect is crucial for anyone looking to use these devices to their full potential, offering a clear path to diagnosing and resolving vehicle issues with confidence.

Reading and Interpreting Codes

When delving into the basic operations of using an automatic diagnostic scanner, a fundamental skill is the ability to read and interpret diagnostic trouble codes (DTCs) that the vehicle's onboard diagnostics (OBD) system generates. These codes are the vehicle's way of communicating specific issues or malfunctions detected in its systems. Understanding these codes is crucial for diagnosing and fixing problems efficiently.

Each code consists of a five-character alphanumeric string, where the first character, a letter, identifies the part of the vehicle's system where the issue was detected. For example, "P" stands for powertrain, which includes the engine and transmission, "B" for body, "C" for chassis, and "U" for user network related to communication between the vehicle's computers.

The second character is a number that indicates whether the code is generic (0) or manufacturer-specific (1). Generic codes are standardized and apply to all vehicle makes and models, while manufacturer-specific codes provide more detailed information for a specific vehicle make.

The third character, a digit, narrows down the part of the system where the issue occurred. For instance, in powertrain codes, numbers like 1 denote fuel and air metering problems, while 3 points to ignition system issues. This specificity helps in targeting diagnostics more accurately.

The last two digits further specify the exact problem within the subsystem identified by the third character. These numbers, ranging from 00 to 99, pinpoint the precise issue, such as a specific sensor failure or misreading.

Reading the codes is the first step, which is relatively straightforward with most scanners automatically retrieving and displaying them upon being connected to the vehicle's OBD port. However, interpreting what these codes actually mean is where the real skill comes in. This requires not only an understanding of what each character in the code represents but also knowledge of how the vehicle's systems interact and how a problem in one area can affect others.

For effective interpretation, it's beneficial to consult a reliable database or manual that lists DTCs and their meanings. Many scanners come with built-in libraries or companion apps that provide detailed definitions and troubleshooting tips for each code. Additionally, understanding the context in which the code has appeared is important—considering factors such as symptoms observed in the vehicle's performance, conditions under which the issue occurs, and any recent repairs or modifications can offer valuable clues for accurate diagnosis.

Once the codes have been interpreted, the next steps involve researching the potential causes and solutions for the identified issues. This might include consulting vehicle-specific repair manuals, online forums, or databases that offer detailed diagnostic procedures and repair strategies based on the codes retrieved.

Ultimately, the goal is to not just read and understand the codes but to use that information to guide effective repair and maintenance decisions. This skill set enables mechanics and vehicle owners alike to address problems promptly and accurately, ensuring the vehicle's longevity and reliability.

Clearing Codes

Clearing codes from a vehicle's onboard diagnostic system is a fundamental operation that can be performed with an automatic diagnostic scanner. This procedure typically follows the identification and resolution of the issues indicated by the trouble codes. It's crucial for users to understand that clearing codes does not fix a problem by itself; it should only be done after the underlying issues have been properly addressed. The process involves connecting the scanner to the vehicle's OBD-II port, usually found under the dashboard, and navigating through the device's menu to find the option to clear or reset the codes.

Before clearing any codes, it's important to read and document them thoroughly, as they provide valuable insights into the vehicle's condition. Many scanners allow users to save or export these codes along with the freeze frame data, which captures the vehicle's operating conditions at the time of the fault. This information can be invaluable for future troubleshooting or if the problem reoccurs.

After ensuring that the issues indicated by the codes have been rectified, the user can proceed to clear them using the scanner. This action resets the vehicle's onboard computer, erasing the codes and turning off the check engine light (CEL) or other warning lights. However, it's important to note that if the underlying issue has not been adequately addressed, the lights and codes will likely reappear after the vehicle has been driven for a certain distance or over a period of time.

Clearing codes also plays a role in vehicle maintenance and inspection routines. In some jurisdictions, vehicles must pass an emissions test or a diagnostic scan to renew registration. Clearing codes without fixing the issue may temporarily turn off warning lights, but the vehicle is likely to fail these tests if the problems persist.

Moreover, modern vehicles often use a drive cycle—a specific set of operating conditions and driving procedures—to reset all monitors to a "ready" state after codes have been cleared. If a vehicle is presented for an emissions test before these monitors are set to "ready," it may fail the test even if there are no active trouble codes.

In addition to technical aspects, ethical considerations come into play when clearing codes. For instance, clearing codes to conceal known issues, especially when selling a vehicle, is deceptive and can lead to legal consequences. It's essential for users to employ diagnostic scanners responsibly, ensuring that the clearing of codes is part of a comprehensive approach to vehicle repair and maintenance.

Lastly, regular use of a diagnostic scanner for monitoring vehicle health and clearing codes, when appropriate, can help maintain the vehicle in optimal condition, improve fuel efficiency, and potentially avoid costly repairs down the line. However, users should always seek professional advice when in doubt about the nature of a problem or the appropriateness of clearing codes, to ensure the safety and reliability of their vehicle.

Advanced Functions

Live Data Monitoring

Live data monitoring is a pivotal function of advanced diagnostic scanners, enabling users to view real-time data from a vehicle's sensors and systems as it operates. This feature is essential for diagnosing intermittent problems that might not trigger a fault code or for verifying that a repair has effectively resolved an issue. By accessing this live information, mechanics and enthusiasts can see exactly what is happening within the vehicle's various systems at any given moment, providing a dynamic snapshot of its health and performance.

The capability to monitor live data transforms diagnostics by allowing for a more nuanced understanding of how a vehicle's engine, transmission, emissions systems, and other critical components interact under different conditions. For instance, live data can show real-time outputs from the oxygen sensors, which play a crucial role in the engine's fuel management system, or provide instant feedback on the effectiveness of the catalytic converter, crucial for emissions control.

Furthermore, live data monitoring facilitates a more targeted approach to troubleshooting. Rather than relying on generic trouble codes, mechanics can observe how changes in one system affect others in real-time, identifying the root cause of complex issues more quickly. This approach is invaluable for diagnosing problems that do not trigger a specific error code, such as intermittent misfires, fluctuating idle speeds, or issues that occur under specific driving conditions.

Advanced diagnostic scanners with live data capabilities often allow users to customize their data displays, enabling them to focus on specific parameters relevant to the diagnosis at hand. This can include graphical representations of sensor outputs, making it easier to spot trends or anomalies over time. For example, a graph showing the engine coolant temperature might reveal overheating issues before they become severe enough to cause damage or trigger a warning light.

Using live data monitoring effectively requires a solid understanding of vehicle systems and how they are supposed to operate under normal conditions. It is here that the diagnostic scanner becomes more than just a tool for reading codes; it becomes an instrument for deepening one's understanding of automotive technology and enhancing diagnostic skills.

Moreover, live data monitoring is not just about diagnosing problems. It's also a powerful way to verify that repairs have been successful. After replacing or repairing components, real-time data can confirm that the vehicle is functioning as expected, providing both the technician and the vehicle owner with peace of mind.

Live data monitoring is a critical function for anyone looking to diagnose and understand modern vehicles more effectively. It offers a window into the ongoing operations of a vehicle's systems, providing the insights needed to diagnose issues accurately, verify repairs, and ensure optimal vehicle performance.

System Tests (e.g., EVAP system, O2 sensor)

In the realm of using an automatic diagnostic scanner, one of the advanced functions that stand out involves conducting system tests, such as those for the EVAP system and the O2 sensor. These tests are essential for pinpointing issues within specific vehicle systems that might not be immediately apparent through standard diagnostic codes.

The EVAP (Evaporative Emission Control System) test is particularly crucial as it checks for leaks in the system that can lead to fuel vapors escaping into the atmosphere, a common cause for failing emissions tests. An automatic diagnostic scanner initiates the EVAP test by commanding the vehicle to perform a self-check, where it pressurizes the EVAP system and monitors it for leaks. This process involves several components, including the purge valve, vent valve, and the charcoal canister. By understanding how to use the scanner to initiate and interpret the results of this test, users can diagnose issues like loose fuel caps, damaged EVAP lines, or faulty valves, which might otherwise go unnoticed.

Similarly, testing the O2 sensor, or oxygen sensor, is another advanced function facilitated by diagnostic scanners. The O2 sensor is pivotal for ensuring that the vehicle's engine runs at optimum efficiency by measuring the oxygen levels in the exhaust gases. This data is used by the engine control unit (ECU) to adjust the air-fuel mixture. A diagnostic scanner can be used to monitor the live data from the O2 sensor in real time, checking for proper operation. Through this test, it's possible to identify sensors that are responding too slowly or not at all, which could lead to poor fuel economy and increased emissions.

What sets these advanced functions apart is not just the ability to read error codes but to actively test and interact with the vehicle's systems, providing a dynamic analysis of its condition. Users can benefit from these functions by gaining a deeper understanding of their vehicle's health, beyond what is visible through basic diagnostics. Learning to effectively perform these system tests requires a good grasp of the scanner's operating instructions and an understanding of the vehicle's specific diagnostic protocols.

Furthermore, the ability to perform these tests empowers users to make more informed decisions regarding repairs and maintenance, potentially avoiding unnecessary replacements and focusing on the actual issues. For instance, replacing an O2 sensor without testing might not resolve an underlying issue with fuel mixture or air intake, leading to repeated failures.

Using an automatic diagnostic scanner to conduct these system tests represents a significant step towards advanced vehicle diagnostics. It equips users with the ability to not only identify but also understand complex vehicle issues, paving the way for accurate repairs and efficient maintenance. This knowledge is invaluable for anyone looking to maintain their vehicle's performance and reliability, making an automatic diagnostic scanner a crucial tool in the arsenal of modern automotive care.

Programming and Coding

Programming and coding with an automatic diagnostic scanner represent some of the most sophisticated capabilities available to automotive professionals and enthusiasts alike. This advanced function goes beyond simple fault diagnosis and resetting warning lights; it delves into modifying vehicle parameters, updating electronic control unit (ECU) software, and even reprogramming keys. These tasks are not only delicate but require a deep understanding of the vehicle's onboard computer systems and the right diagnostic tools that offer such capabilities.

Many modern vehicles come equipped with ECUs that control everything from engine timing and fuel injection to the operation of air conditioning and the deployment of airbags. Each of these systems can be fine-tuned or updated through programming and coding. This process often begins with a diagnostic scanner establishing a secure connection to the vehicle's OBD system, ensuring compatibility and a stable link for data transmission.

One common application of programming is the installation of updates provided by the vehicle manufacturer. These updates may enhance vehicle performance, fix known bugs, or improve fuel efficiency. The process is akin to installing a new operating system on a computer but requires precise execution to prevent issues. Another application is coding new keys or reprogramming existing ones, a task that has become more complex as key fobs now often include features like remote start and security encryption.

The capability to modify vehicle parameters allows for customization that can optimize vehicle performance for specific conditions or preferences. For example, the diagnostic scanner can adjust the settings for the suspension system in a performance car for a smoother ride or better handling. However, this level of customization should always be approached with caution, as unauthorized modifications can void warranties or, worse, impair vehicle safety systems.

A critical aspect of programming and coding is the need for up-to-date software on both the diagnostic scanner and the vehicle. Manufacturers regularly release software updates for their diagnostic tools, ensuring they remain compatible with new vehicle models and their software. These updates often include new functionalities, expanded vehicle coverage, and improved user interfaces.

Safety is paramount when performing these advanced functions. Incorrect coding or programming can lead to vehicle malfunctions, compromised safety systems, or even render a vehicle inoperable. It's crucial that users follow the manufacturer's guidelines, use the diagnostic scanner as intended, and understand the impact of the changes they are making. For the most complex tasks, it may be advisable to seek professional assistance, especially when dealing with critical systems like airbag deployment or engine control modules.

Programming and coding through an automatic diagnostic scanner unlock a new realm of possibilities for vehicle diagnostics, maintenance, and customization. These advanced functions, while powerful, demand a comprehensive understanding of vehicle systems, meticulous attention to detail, and a responsible approach to modifying vehicle software. As vehicles continue to evolve, becoming more software-driven, the importance and utility of these functions within diagnostic scanners are set to grow, making them indispensable tools in the automotive industry.

Special Functions (e.g., DPF regeneration, TPMS reset)

Special functions in automatic diagnostic scanners offer advanced capabilities beyond basic fault code reading, providing users with tools to perform specific maintenance and repair tasks that were once exclusive to professional service centers. These functions include but are not limited to Diesel Particulate Filter (DPF) regeneration and Tire Pressure Monitoring System (TPMS) reset, among others.

DPF regeneration is a crucial process for diesel-engine vehicles equipped with a DPF to reduce emissions and prevent the filter from becoming clogged with soot. Over time, the DPF accumulates particulate matter, leading to reduced performance and potentially significant repair costs if not properly maintained. Automatic diagnostic scanners with the DPF regeneration function can initiate a forced regeneration cycle, burning off accumulated particulate matter at high temperatures and restoring the filter's functionality. This process is essential for maintaining the vehicle's fuel efficiency, reducing harmful emissions, and extending the lifespan of the DPF.

The TPMS reset function is equally important for maintaining vehicle safety and efficiency. The TPMS monitors tire pressure, alerting the driver when a tire is significantly under-inflated. Correct tire pressure is critical for ensuring optimal vehicle handling, fuel efficiency, and tire lifespan. After repairing or replacing a tire or TPMS sensor, the system must be reset or relearned to recognize the new or repaired tire. An automatic diagnostic scanner with TPMS reset capability allows users to complete this process quickly, ensuring the TPMS continues to monitor tire pressures accurately without the need for a professional service appointment.

In addition to these, automatic diagnostic scanners offer a variety of other special functions tailored to the specific needs of modern vehicles. These may include battery registration for vehicles with energy management systems that require new batteries to be registered with the vehicle's computer, steering angle sensor calibration which is critical after wheel alignment or suspension repairs, and key programming for vehicles with keyless entry and push-button start systems.

Understanding how to use these advanced functions requires a comprehensive grasp of the vehicle's systems and the specific procedures outlined in the diagnostic scanner's manual. Proper use of these functions can save time and money by avoiding unnecessary trips to the dealership or repair shop, and can also provide insights into the vehicle's overall health and maintenance needs.

However, it is crucial for users to exercise caution and adhere to safety guidelines when performing these advanced functions. Incorrect use can lead to vehicle damage or compromise safety features. Always consult the vehicle's service manual and follow the step-by-step instructions provided by the scanner to ensure that these tasks are performed correctly. Additionally, users should be aware of the legal and warranty implications of performing certain maintenance tasks independently.

By offering these special functions, automatic diagnostic scanners empower vehicle owners and independent mechanics to perform a range of maintenance and repair tasks with precision and confidence, ultimately improving vehicle performance, extending vehicle life, and enhancing driving safety.

Troubleshooting Common Issues

Connectivity Problems

When dealing with automatic diagnostic scanners, one of the most common hurdles encountered by users is connectivity problems. These issues can manifest in various forms, such as the scanner failing to establish a connection with the vehicle's onboard diagnostics system, intermittent connections, or the scanner not being recognized by the accompanying software on a computer or mobile device. Several factors can contribute to these problems, ranging from simple fixes to more complex issues requiring detailed troubleshooting.

Firstly, a fundamental step in resolving connectivity issues involves checking the physical connections. This includes ensuring that the diagnostic scanner's cable is securely plugged into the vehicle's OBD-II port, which is typically located under the dashboard near the steering column. For wireless scanners, it's crucial to verify that the Bluetooth or Wi-Fi connection is properly configured and that the device is within range.

Another common issue pertains to compatibility. Diagnostic scanners are designed to work with a wide range of vehicle makes and models, but not all scanners are universal. It's important to confirm that the scanner supports the specific vehicle you are attempting to diagnose. Compatibility lists are often provided by the scanner's manufacturer.

Software and firmware updates represent another critical area in the troubleshooting process. Manufacturers frequently release updates to enhance scanner functionality, add new features, or resolve known bugs. Ensuring that the scanner's firmware and any associated software are up to date can often resolve connectivity issues.

Vehicle-specific issues can also impede connectivity. Some vehicles may have unique OBD-II protocols or security features that require specific steps to establish a connection. In such cases, consulting the vehicle's manual or a professional database for guidance can provide vehicle-specific troubleshooting steps.

Power issues within the vehicle itself can lead to connectivity problems. The OBD-II port is powered by the vehicle's battery, so a weak or dead battery may not provide sufficient power for the scanner to operate. Checking the vehicle's battery health and ensuring the ignition is in the correct position, usually the "On" or a similar setting, can help resolve these issues.

In instances where the diagnostic scanner is not recognized by a computer or mobile device, checking the device manager (on PCs) or the settings menu (on mobile devices) to ensure the scanner is detected is a good step. Sometimes, manual selection of the connection or updating device drivers can rectify the problem.

Environmental factors can also play a role, especially for wireless connections. Interference from other wireless devices, physical obstructions, and even weather conditions can affect connectivity.

Minimizing interference by changing locations or disabling other wireless devices temporarily can help establish a stable connection.

Finally, if all else fails, resetting the diagnostic scanner to its factory settings can resolve underlying issues that are not immediately apparent. This step should be taken with caution, as it may erase any saved data or custom settings on the device.

Troubleshooting connectivity problems with diagnostic scanners involves a systematic approach, starting with the simplest solutions and moving towards more complex diagnostics. While these issues can be frustrating, patience and a methodical process often lead to resolving the problem, allowing for successful vehicle diagnostics.

Software and Update Issues

Software and update issues are among the most common challenges faced by users of automatic diagnostic scanners. These problems can range from difficulties in connecting the scanner to a vehicle's onboard diagnostics system to errors in the scanner's operation due to outdated software. Understanding how to troubleshoot these issues is crucial for ensuring that the diagnostic scanner functions effectively and provides accurate data.

When a diagnostic scanner fails to connect to a vehicle, it's often due to compatibility issues between the scanner's software and the vehicle's firmware. This is particularly common with newer vehicles or after a vehicle has received a firmware update. To resolve this, users should ensure that their diagnostic scanner is updated with the latest software version available from the manufacturer. Most manufacturers offer free software updates through their websites, which can be downloaded and installed using a computer.

Another common issue arises when the diagnostic scanner operates slowly or with errors. This can be due to corrupted software or incomplete updates. In such cases, performing a factory reset of the scanner and then reinstalling the latest software version can often resolve the problem. It's important to follow the manufacturer's instructions carefully during this process to avoid further issues.

Software updates not only resolve compatibility and performance issues but also often add new features and functionalities to the diagnostic scanner, such as support for additional vehicle models or enhanced diagnostic capabilities. Therefore, regularly checking for and installing software updates is an essential maintenance task for any diagnostic scanner owner.

However, updating the software can sometimes lead to unexpected issues, such as the scanner failing to start up or losing stored diagnostic data. Before updating the scanner's software, it's advisable to back up any important data. If an update does cause issues, consulting the manufacturer's support documentation or contacting their customer service can provide guidance on how to resolve the problem. In some cases, it may be necessary to reinstall an older version of the software until a new update is available that resolves the issue.

Finally, users should be mindful of the source of their software updates. Downloading updates from unofficial or unauthorized sources can lead to the installation of malicious software or software that is incompatible with the scanner, potentially causing irreparable damage. Always use the official website or contact the manufacturer directly to obtain software updates.

By understanding how to troubleshoot software and update issues, users of automatic diagnostic scanners can ensure that their device remains a reliable tool for vehicle diagnostics. Regular maintenance, including software updates, is key to maximizing the scanner's performance and longevity.

Interpreting Complex Codes

Interpreting complex codes from an automatic diagnostic scanner requires a blend of technical knowledge, critical thinking, and sometimes a bit of detective work. When a diagnostic scanner throws a code that isn't straightforward, the challenge isn't just understanding what the code means but also deciphering the underlying issue it points to. Complex codes often indicate problems that are intermittent or influenced by a combination of failing components rather than a single malfunction.

The first step in interpreting complex codes is to consult the vehicle's service manual alongside the diagnostic tool's database. These resources often provide a more detailed explanation of what each code means, including possible causes and the specific systems involved. It's essential to remember that a code by itself does not directly indicate which part needs to be replaced. Instead, it suggests which system is experiencing the problem, guiding the technician to further investigation.

Once the meaning of the code is understood, the next step involves looking at the live data and freeze frame data provided by the scanner. Live data allows for real-time monitoring of the vehicle's systems while operating, offering clues that static codes may not reveal. Freeze frame data captures the vehicle's operating conditions at the moment the fault occurred, providing valuable context that can help pinpoint intermittent issues.

Understanding the relationships between different vehicle systems is crucial when dealing with complex codes. For instance, a problem in the fuel delivery system might throw codes related to the engine's performance, while the root cause could be a failing fuel pump or clogged fuel filter. Similarly, a misfiring engine could result from various factors, including ignition system failures, fuel system problems, or even a malfunctioning sensor affecting air-fuel mixture.

Another key strategy is to research common issues with the specific make and model of the vehicle. Online forums, technical bulletins, and manufacturer advisories can provide insights into recurring problems, which might be related to the complex code being investigated. This approach can save time by directing the technician to known issues that match the symptoms and codes displayed by the diagnostic scanner.

Advanced diagnostic procedures may be required for some complex codes. This could involve using additional diagnostic tools like oscilloscopes to measure electrical signals, or performing manual tests such as compression or leak-down tests to assess engine health. In some cases, it's necessary to methodically test individual components within the implicated system to rule out potential causes.

Documenting each step of the diagnostic process is crucial, especially when dealing with intermittent issues that might not be immediately reproducible. Notes on conditions, tests performed, and observations can be invaluable for tracking down elusive problems.

Ultimately, interpreting complex codes is about piecing together a puzzle where each piece of data helps form a clearer picture of the vehicle's condition. It requires patience, a methodical approach, and an understanding that sometimes, the issue indicated by a complex code can be the symptom of a larger problem, requiring comprehensive investigation to ensure accurate diagnosis and repair.

Addressing Inaccurate or Inconsistent Data

When using an automatic diagnostic scanner, encountering inaccurate or inconsistent data can be perplexing and hinder the diagnostic process. Such issues often arise due to a variety of factors, including but not limited to, faulty sensors, poor connections, interference, outdated software, or even the scanner's compatibility with the vehicle's onboard diagnostic system. To effectively address these challenges, a methodical approach is required.

Firstly, ensuring that the diagnostic scanner is properly connected to the vehicle's diagnostic port is fundamental. A loose or poor connection can result in intermittent data transmission, leading to inconsistencies in the readings. It's also essential to inspect the condition of the port and the scanner's connector for any physical damage or dirt that might interfere with the connection.

Another common source of inaccurate data is faulty sensors within the vehicle. Sensors are the primary source of the data fed into the onboard diagnostics system. If a sensor is malfunctioning or has failed, it will send incorrect information to the scanner. Testing sensors individually for proper operation can help isolate and correct the source of inaccurate data.

Electrical interference within the vehicle's electrical system can also lead to erroneous scanner readings. Sources of such interference can include damaged wiring, poor grounding, or aftermarket electronic devices connected to the vehicle. Inspecting and repairing the vehicle's electrical system can mitigate these issues.

Software-related issues are another significant factor that can affect the accuracy of diagnostic data. Ensuring that the scanner's software is up to date is crucial since updates often include fixes for known bugs, improvements in data interpretation algorithms, and expanded vehicle coverage. Similarly, the vehicle's onboard computer software should be up to date, as manufacturers release updates that can improve the system's functionality and communication clarity.

Compatibility between the scanner and the vehicle's specific model and year can influence the accuracy of the diagnostic data. Not all scanners are universal, and some may not fully support all the features or protocols used by a particular vehicle's onboard diagnostics system. Consulting the scanner's compatibility list and using a scanner recommended or approved by the vehicle's manufacturer can alleviate compatibility issues.

When faced with persistent inaccurate or inconsistent data, referring to additional sources of information can be invaluable. Technical service bulletins (TSBs) issued by vehicle manufacturers often address known issues with specific models, including problems that may affect diagnostic data. Professional forums and communities can also be a resource for insights from experienced technicians who may have encountered similar issues.

In cases where standard troubleshooting does not resolve the issue, seeking assistance from the scanner's manufacturer or a professional with specialized knowledge in automotive diagnostics may be necessary. They can provide support and insights based on a deep understanding of diagnostic technologies and vehicle electronics.

Ultimately, addressing inaccurate or inconsistent data from a diagnostic scanner involves a combination of thorough connection and sensor checks, ensuring up-to-date software, considering compatibility, and seeking further information or assistance when needed. By systematically approaching these issues, users can enhance the reliability and accuracy of their diagnostic efforts, leading to more effective vehicle maintenance and repair.

Maintenance and Care of Your Diagnostic Scanner

Regular Software Updates

Regular software updates are vital for the optimal performance and longevity of your automatic diagnostic scanner. These updates are designed to enhance the functionality of the device, add new features, and expand the range of vehicles and systems it can diagnose. As automotive manufacturers release new models and update the software on existing vehicles, diagnostic scanners must be updated to keep pace with these changes. Without regular updates, a scanner might not recognize the latest models or might display inaccurate or incomplete data, leading to misdiagnosis.

Manufacturers of diagnostic scanners frequently release software updates to correct bugs, improve user interface experiences, and expand the code library—the comprehensive database that the scanner uses to interpret diagnostic codes from a vehicle. This library is constantly evolving as new diagnostic codes and vehicle systems are developed. An up-to-date code library is essential for accurate diagnostics and troubleshooting.

Another critical aspect of software updates is the enhancement of security features. As scanners become more sophisticated, often connecting to the internet for updates or to access cloud-based resources, they become vulnerable to cybersecurity threats. Updates can patch security vulnerabilities, protecting both the scanner and the data it accesses from unauthorized access or malicious attacks.

To ensure your diagnostic scanner remains effective, it's crucial to establish a routine for checking and applying software updates. This might involve subscribing to notifications from the manufacturer, regularly visiting their website, or using an automatic update feature if your scanner is equipped with one. While some updates are free, others, especially those that significantly expand the scanner's capabilities or update libraries for new vehicle models, might require a subscription or a one-time purchase.

Incorporating regular software updates into the maintenance routine for your diagnostic scanner not only enhances its performance and reliability but also protects your investment. As vehicles become more complex, the ability of your scanner to accurately diagnose and resolve issues is paramount. Staying current with software updates ensures that you can continue to offer or benefit from the highest level of diagnostic accuracy and efficiency.

Hardware Maintenance

Maintaining the hardware of an automatic diagnostic scanner is crucial for ensuring its longevity and reliability. Regular hardware maintenance not only prevents the device from physical damage but also ensures accurate diagnostics and uninterrupted service. The first step in hardware maintenance involves routinely inspecting the scanner for any visible signs of wear or damage. This includes checking the device's casing for cracks, the integrity of the connection ports, and the condition of the cables and connectors, as these are often subjected to wear and tear through regular use.

Keeping the scanner clean is essential. Dust, dirt, and automotive fluids can accumulate on the device and its connectors, potentially leading to poor connections or device failure. Use a soft, dry cloth to wipe down the scanner and its components. For tougher grime, a cloth slightly dampened with isopropyl alcohol can be used, but it's important to ensure that no liquid enters the device.

The screen of the scanner, if applicable, is one of its most delicate parts and should be cleaned with care. Special screen cleaners or a soft, lint-free cloth slightly dampened with water can be used to gently clean the screen. Avoid harsh cleaning agents or abrasive cloths that could scratch or damage the screen.

Regularly checking and updating the software of the diagnostic scanner is also a part of hardware maintenance, as software updates often include firmware upgrades that enhance the hardware functionality or longevity. This proactive approach can prevent software issues that may manifest as hardware problems, such as unresponsive buttons or slow performance.

The connectors and ports are critical for the functionality of the scanner, as they facilitate the connection between the scanner and the vehicle. Keeping these connectors clean and free from debris is essential. A can of compressed air can be used to blow out any dust or debris from the ports, and electrical contact cleaner can help maintain a good connection.

Storage plays a significant role in hardware maintenance. When not in use, the scanner should be stored in a clean, dry place, away from direct sunlight and extreme temperatures, which can damage the device's internal components. Many scanners come with a protective case or bag, which provides an ideal storage solution, protecting the device from physical damage and dust.

Battery maintenance is applicable for scanners that use rechargeable batteries. Ensuring that the battery is charged and stored according to the manufacturer's recommendations can prevent battery-related issues and extend the battery's life. It is also important to monitor the health of the battery and replace it when its ability to hold a charge diminishes.

Finally, considering the physical handling of the scanner can significantly impact its durability. Gentle handling, avoiding drops or rough treatment, and using the device in accordance with the manufacturer's guidelines can prevent many common forms of hardware damage.

By adhering to these maintenance practices, users can significantly extend the life of their automatic diagnostic scanners, ensuring they remain a reliable tool for vehicle diagnostics for years to come.

Storage and Handling

Proper storage and handling of an automatic diagnostic scanner are essential to ensure its longevity and reliability. These devices, sophisticated and delicate, require a careful approach to maintain their functionality over time. Storing the scanner in a clean, dry, and temperature-controlled environment is paramount. Exposure to extreme temperatures, moisture, or dust can lead to hardware malfunction or degrade sensitive electronic components. Ideally, the scanner should be kept in a protective case when not in use, which shields it from accidental drops, dust, and direct sunlight, all of which can be detrimental to its internal and external components.

Handling the scanner with care is equally important. Before use, ensure that your hands are clean and dry to prevent any dirt or moisture from getting into the device. When connecting or disconnecting it from a vehicle, do so gently to avoid damaging the connector pins or the vehicle's diagnostic port. These ports and connectors are the primary interfaces for the scanner and can suffer from wear and tear if not treated with care.

Regularly cleaning the scanner is also crucial for its upkeep. Use a soft, dry cloth to wipe the exterior surfaces of the device, avoiding harsh chemicals or abrasive materials that could damage the screen or casing. For the screen, a microfiber cloth is recommended to prevent scratching. If the scanner has any ports or openings, they should be cleaned using compressed air to remove dust or debris that could interfere with the connections.

Software updates are a critical aspect of scanner maintenance. Manufacturers often release updates to improve functionality, add new features, or address known bugs. Regularly updating the scanner's software ensures it operates efficiently and remains compatible with newer vehicle models and technologies. Always follow the manufacturer's instructions for updates closely to avoid installation errors that could impair the device.

Battery maintenance is another key consideration, especially for portable models. Rechargeable batteries should be kept charged and used regularly to prevent degradation. If the scanner will not be used for an extended period, the battery should be charged to around 50% to minimize the risk of damage. For scanners with replaceable batteries, ensure that the battery contacts are clean and free from corrosion.

Finally, it's important to review and understand the manufacturer's warranty and service information. Knowing what the warranty covers and how to access repair services can save time and money should the scanner require professional attention. Keeping a record of the purchase date, warranty period, and any services performed on the device can also be helpful for future reference.

By following these storage and handling practices, users can significantly extend the life of their automatic diagnostic scanner, ensuring it remains a reliable tool for vehicle diagnostics for years to come.

Conclusion

Wrapping up the journey through the intricacies of using an automatic diagnostic scanner, it becomes evident that mastering this tool is not just about leveraging technology but about embracing a proactive approach to vehicle maintenance and repair. The guide has meticulously laid out the pathways from understanding the basics to navigating the complexities of diagnostic codes, enabling users to unlock the full potential of their vehicles. This knowledge does more than just save time and money; it empowers individuals, fosters a deeper connection with the machinery, and demystifies the digital dialogue between man and machine.

Through the chapters, readers have been equipped with the skills to not only perform diagnostics but to interpret and act on the information provided by their scanners. This elevates the diagnostic scanner from a mere troubleshooting tool to an essential instrument of insight into the health and performance of vehicles. It underscores the significance of staying abreast with technological advancements and adapting to the evolving landscape of automotive diagnostics.

Moreover, the emphasis on proper storage, handling, and maintenance of the scanner underlines the importance of treating these devices as valuable investments. By caring for them, users ensure their longevity and reliability, making them indispensable allies in the quest for vehicle longevity and safety.

In the grand scheme, the journey with the diagnostic scanner is about more than just diagnosing problems; it's about embracing a culture of prevention, education, and empowerment. As vehicles become increasingly sophisticated, the ability to speak their language through a diagnostic scanner becomes an invaluable skill, bridging the gap between the past and the future of automotive repair.

Ultimately, the guide on "How to Use an Automatic Diagnostic Scanner" serves as a beacon for all those navigating the ever-expanding sea of automotive technology. It stands as a testament to the power of knowledge, the importance of preparedness, and the endless possibilities that open up when we choose to understand the heartbeats of the machines that move us. As readers turn the final page, they are not at the end of a chapter but at the beginning of a journey, armed with the knowledge and confidence to face whatever lies on the road ahead.

www.ingramcontent.com/pod-product-compliance
Lightning Source LLC
Chambersburg PA
CBHW050013230526
45470CB00003B/944